FINISH THE PATTERN

A TURN-AND-SEE BOOK

by Cari Meister

PEBBLE
a capstone imprint

Patterns are things that repeat
or happen in a certain way.
They are predictable. You know
what's going to happen next,
like the days of the week.

Use clues from the text and the
photos to guess what comes
next. Then turn the page and
see if you are right!

Time for a cold treat on a
hot day. Look at those fruity
popsicles! Red, orange, yellow,
red, orange . . .

What comes next?

Yellow! What a delicious
way to cool off!

It's time for a game of hopscotch!
First, hop on one leg. Then, hop on
two. One leg, two legs, one leg . . .

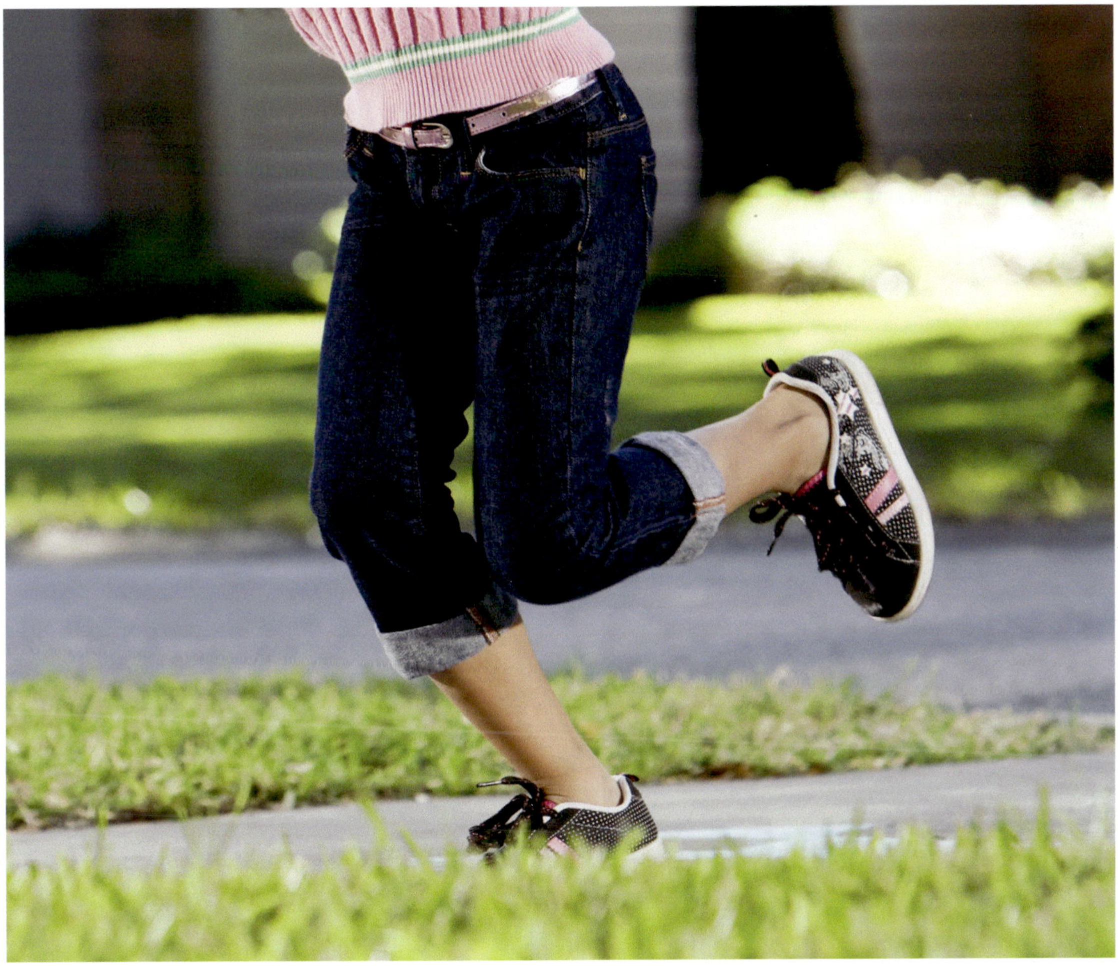

What comes next?

Two legs! Keep on hopping!

Every season brings new
adventures. Spring, summer,
fall, and . . .

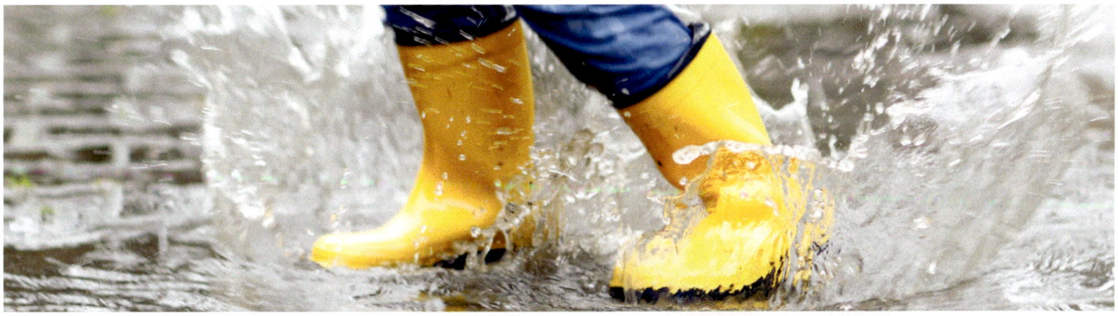

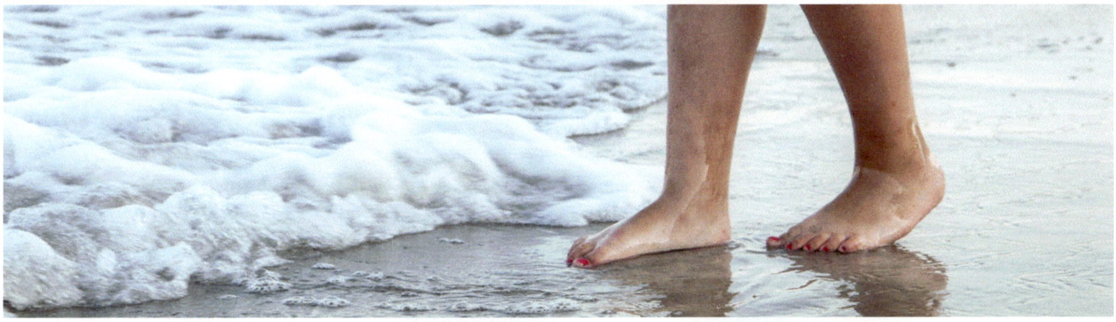

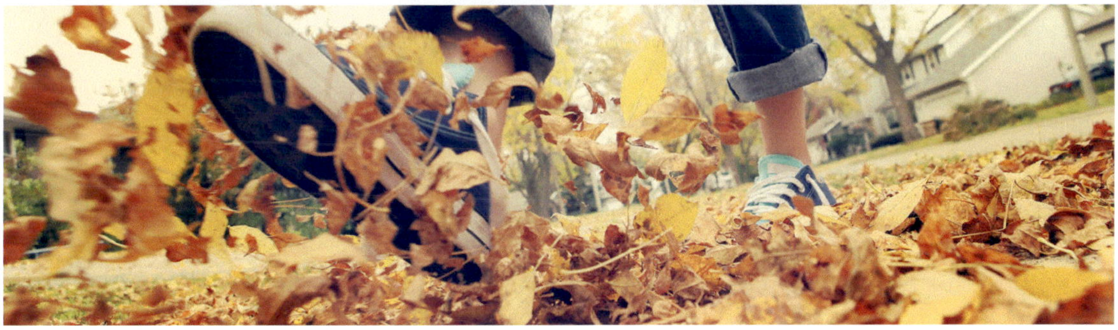

What comes next?

Winter! Brrr!

Yummy donuts lined up in a row.
Chocolate, strawberry, chocolate,
strawberry . . .

What comes next?

Chocolate! Sprinkles and frosting
can't be beat!

Tacey adds one block to each of her towers. One, two, three, four, five . . .

What comes next?

Six! Her purple tower
has six blocks.

There are seven days in a week.
Sunday, Monday, Tuesday,
Wednesday, Thursday, Friday, and . . .

What comes next?

Saturday! Saturday is laundry day.

January, February, March, April, May, June, July, August, September, October, November, and December are the 12 months of the year. Grandma's birthday is in June. Denzel's birthday is the next month.

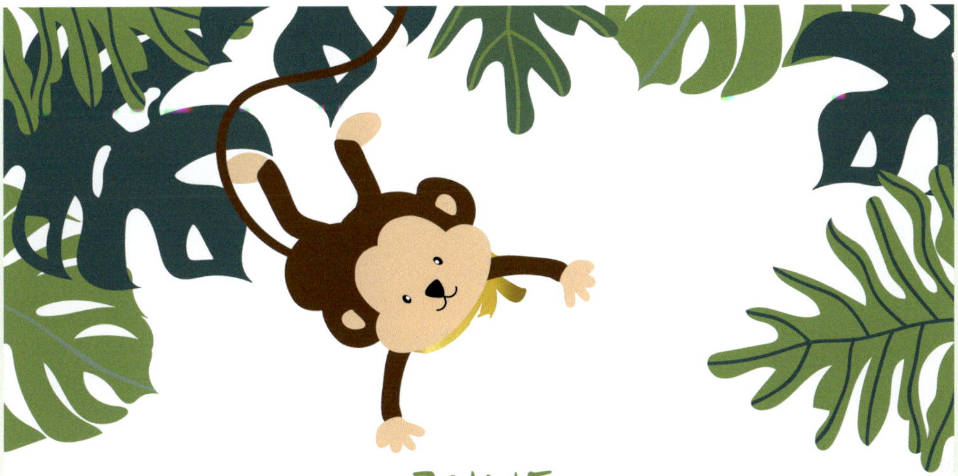

JUNE

Su	Mo	Tu	We	Th	Fr	Sa
	1	2	3	4	5	6
7	8	9	10	11	12	13
14	15	16	17	18	19	20
21	22	23	24	25	26	27
28	29	30				

What month comes next?

turn and see

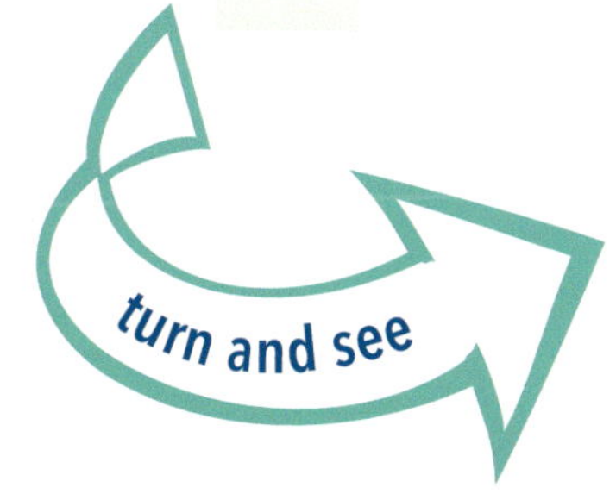

July! Happy birthday, Denzel!

Nolan just learned how to count
by 2s. He put out his dominoes
and started counting the dots.
Two, four, six, eight, ten . . .

What comes next?

12! Look at all those dots!

Julia and her dad are making a colorful salad. It's time to add some peppers! Yellow, red, yellow, red . . .

What comes next?

Yellow! Look at all those vegetables!

Declan lined up his cards and
counted from highest to lowest.
Nine, eight, seven, six, five, four . . .

What comes next?

Three! Declan has the three of diamonds.

Apples are Penelope's favorite
fruit. They are crunchy and sweet!
Red, green, red, green, red . . .

What comes next?

Green! What a tasty afternoon snack!

You have the perfect pumpkin to
carve. First, draw a face. Next,
pick out all the seeds.

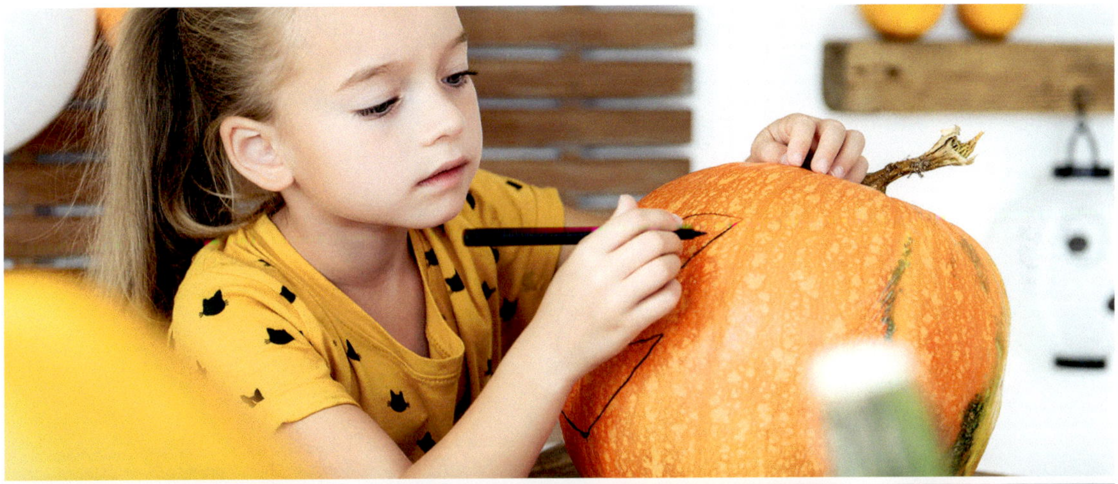

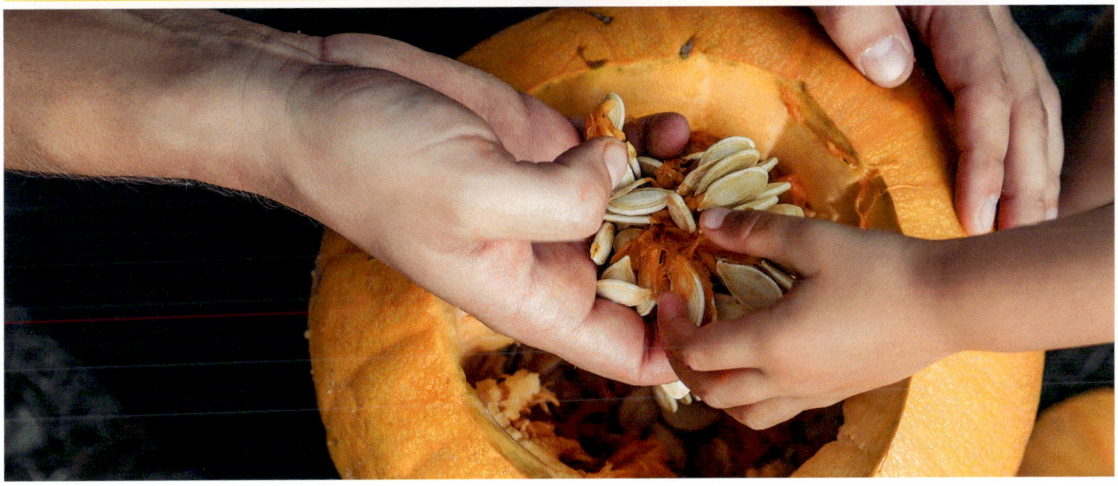

What do you do last?

You carve it!

Let's build a snowman! First,
roll some sticky snow into three
balls. Next, stack the balls.

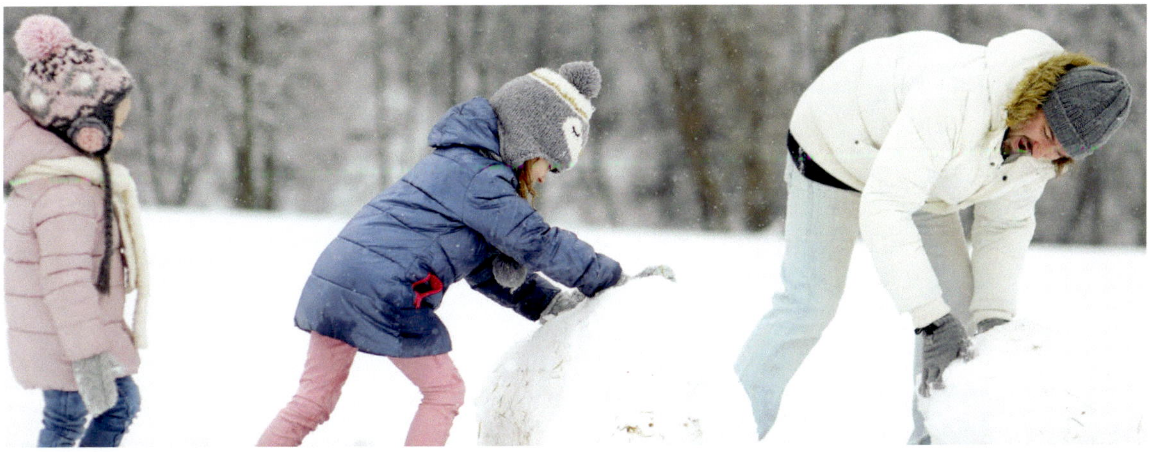

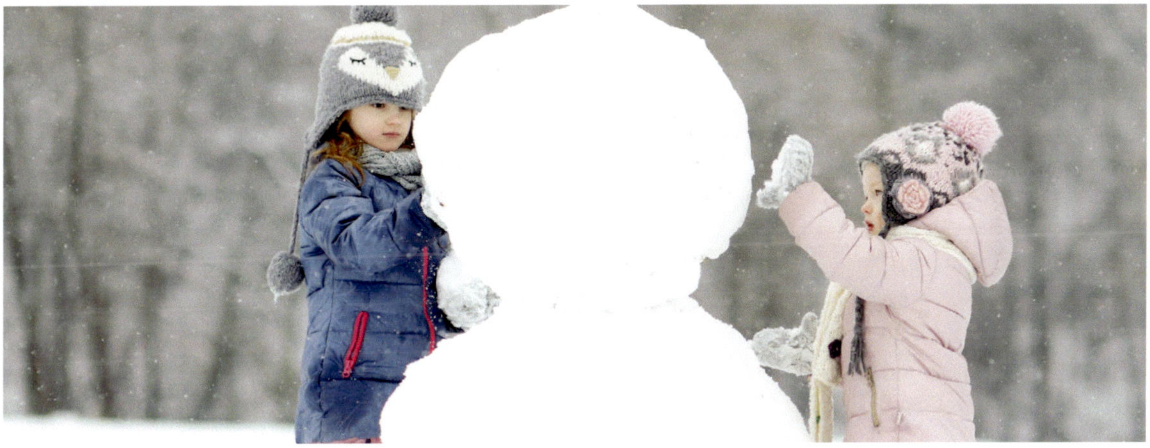

What happens last?

You give your snowman
stick arms and a big smile!

Friday night is pizza night! First,
roll the dough. Next, spread
on the sauce. Then, smother it
with cheese and toppings.

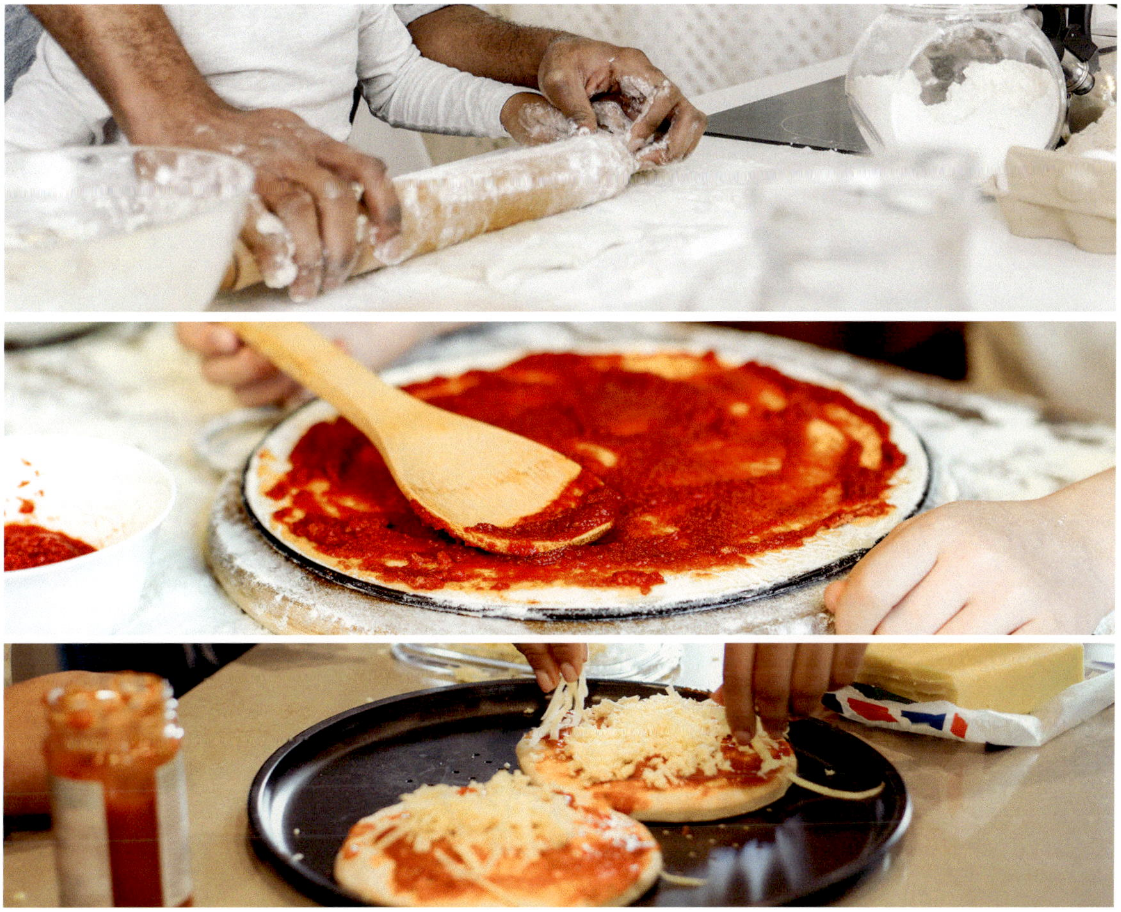

What happens last?

You bake it and eat it!

PATTERNS REVIEW

Color Patterns: popsicles, peppers, apples

Flavor Pattern: donuts

Number Patterns: hopscotch, blocks, dominoes, cards

Seasons: spring, summer, fall, winter

Days of the Week: Sunday, Monday, Tuesday, Wednesday, Thursday, Friday, Saturday

Months: January, February, March, April, May, June, July, August, September, October, November, December

First, Next, Last: carving a pumpkin, building a snowman, making a pizza

Pebble Sprout is published by Pebble, an imprint of Capstone.
1710 Roe Crest Drive
North Mankato, Minnesota 56003
www.capstonepub.com

Library of Congress Cataloging-in-Publication Data
Names: Meister, Cari, author.
Title: Finish the pattern : a turn-and-see book / by Cari Meister.
Description: North Mankato : Pebble, 2021. | Series: Pebble Sprout:
What's next? |Audience: Ages 6-8 | Audience: Grades 2-3 | Summary:
"From the days ofthe week to playing hopscotch, patterns are all around
you. Keep turningthe pages to find out how your life is full of patterns and
sequences"— Provided by publisher.
Identifiers: LCCN 2020037991 (print) | LCCN 2020037992 (ebook) |
ISBN 9781977131553 (hardcover) | ISBN 9781977154224 (pdf) | ISBN
9781977155931 (kindle edition)
Subjects: LCSH: Sequences (Mathematics)—Juvenile literature. | Pattern
perception—Juvenile literature.
Classification: LCC QA292 .M45 2021 (print) | LCC QA292 (ebook) |
DDC 003/.1—dc23
LC record available at https://lccn.loc.gov/2020037991
LC ebook record available at https://lccn.loc.gov/2020037992

Editorial Credits
Editor: Christianne Jones; Designer: Tracy McCabe; Media Researcher: Morgan
Walters; Production Specialist: Kathy McColley